Growing Goods in a Growing Country

Contents

What's Growing?

What are you having for breakfast? Maybe pancakes? Maybe oatmeal with strawberries? A glass of orange juice? Where does the food we eat come from?

The fruits and vegetables we eat come from many places. Many food crops are grown in the United States. Crops such as cotton and sunflowers are also grown here. There are ten farming regions that add to the crops grown in the United States.

United States Agricultural Regions

PACIFIC
NORTHERN PLAINS
LAKE REGION
NORTHEAST
MOUNTAIN
CORN BELT
APPALACHIA
DELTA REGION
SOUTHEAST
SOUTHERN PLAINS
ATLANTIC OCEAN
Gulf of Mexico

N
W   E
S

0   250   500 miles
0   250   500 kilometers

PACIFIC
PACIFIC

Did You Know?

The United States exports more agricultural products than it imports.

The Northeast Region

Climate and soil help farmers decide what crops to grow in each area. Climate includes temperatures, rain, and wind. Soil is the top layer of Earth where plants grow.

The soil and climate in the Northeast are good for growing fruits and vegetables. Grains like wheat, corn, oats, and barley grow in this area.

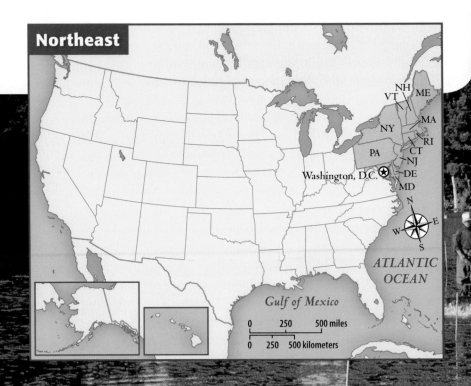

Northeast Region Crops	
State	**Crop**
Delaware	Soybeans
Maine	Potatoes
Rhode Island	Corn
Massachusetts	Cranberries

Pennsylvania is the leading **agricultural** producer of the Northeast region. It ranks 20th in the country. The table below lists some of this state's top crops.

Pennsylvania's Top Crops (2005)		
Crop	Acres Harvested	Yield
Grains and Legumes		
Hay	1,600,000	2.12 tons
Corn for grain	960,000	122 bushels
Vegetables		
Sweet corn	17,700	3.05 tons
Potatoes	11,000	12.50 tons
Fruit		
Apples	21,800	11.80 tons
Grapes	12,000	7.50 tons

Cranberries shown at harvest. It takes 75–100 days for cranberries to turn dark red.

The Corn Belt Region

The Corn Belt has good soil. The states in this area have climates for growing grains and **legumes**. Three of the states (Illinois, Indiana, and Ohio) are top producers of soybeans and corn.

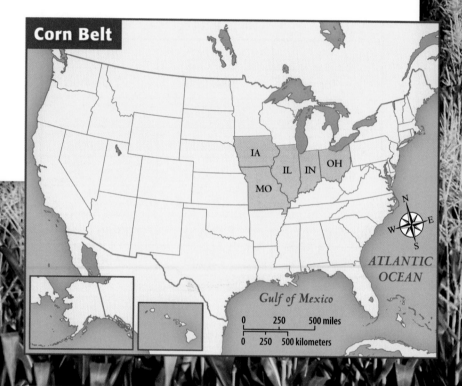

Corn Belt

Interesting Soybean Facts
- During the Civil War, soybeans were used in place of coffee.
- One acre of soybeans can make 82,368 crayons.
- Farmers first grew soybeans as cattle feed.
- $\frac{1}{2}$ of all daily U.S. newspapers are printed with soybean oil–based ink.

Iowa is the leading agricultural producer of the Corn Belt states. 88% of the land in Iowa is farm land. Look at the chart to see some of this state's important crops.

Iowa's Top Crops (2005)		
Crop	Acres Harvested	Yield
Grains and Legumes		
Corn for grain	12,500,000	173 bushels
Soybeans	10,050,000	53 bushels
Hay	1,600,000	3.7 tons
Oats	125,000	79 bushels
Wheat	15,000	50 bushels

Soybeans in pods

Corn fields are common in the Corn Belt.

The Great Lakes Region

The soil and climate in the Great Lakes region states are good for growing grains, fruit, and vegetables.

Great Lakes Region

Great Lakes Region Crops	
State	**Crop**
Wisconsin	Cranberries and cherries
Michigan	Apples, asparagus, blueberries, cherries, and potatoes
Minnesota	Grains

Minnesota is the leading agricultural producer of the Great Lakes region. It has about 19,600 farms. Look at the chart to see some of this state's top crops.

Minnesota's Top Crops (2005)		
Crop	Acres Harvested	Yield
Grains and Legumes		
Corn for grain	6,850,000	174 bushels
Soybeans	6,800,000	45 bushels
Vegetables		
Sugar beets	460,000	20.4 tons
Potatoes	43,000	20.5 tons

Did You Know?

In the United States, corn production measures more than 2 times that of any other crop.

The Appalachian Region

Two important crops of the Appalachian region are peanuts and cotton. Vegetable crops include sweet potatoes and cabbage. A popular fruit grown in the Appalachian region is apples.

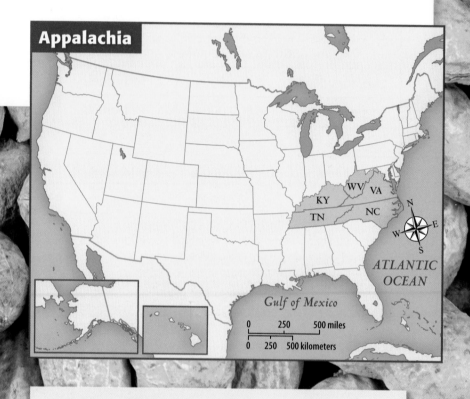

Appalachia

WV VA
KY
TN NC

N
W ✦ E
S

ATLANTIC OCEAN

Gulf of Mexico

0   250   500 miles
0 250 500 kilometers

Fun Facts about Peanuts
- Peanuts make up $\frac{2}{3}$ of all snack nuts eaten in the United States.
- The average American eats more than 6 pounds of peanuts and peanut butter each year.
- Dr. George Washington Carver studied and created more than 300 uses for peanuts.

North Carolina is the leading agricultural producer of the Appalachian region. It ranks 2nd in production of cucumbers for pickles. Look at the chart to see some of this state's top crops.

Fun Facts about Cotton

- A bale of cotton weighs about 480 pounds.
- American paper money is $\frac{3}{4}$ cotton lint and $\frac{1}{4}$ linen.
- A bale of cotton can be made into 313,600 dollar bills.

North Carolina's Top Crops (2005)		
Crop	**Acres Harvested**	**Yield**
Grains and Legumes		
Soybeans	1,460,000	27 bushels
Corn for grain	700,000	120 bushels
Vegetables		
Sweet potatoes	35,000	8.5 tons
Cucumbers	16,000	4.3 tons
Other		
Cotton	810,000	847 pounds
Peanuts	96,000	1.5 tons

The Southeast Region

The Southeast is known for **citrus** fruit. Florida produces more oranges, grapefruit, and tangerines than any other state. Florida is second (to California) in growing vegetables. The Southeast also grows the 3 Ps: peanuts, pecans, and peaches.

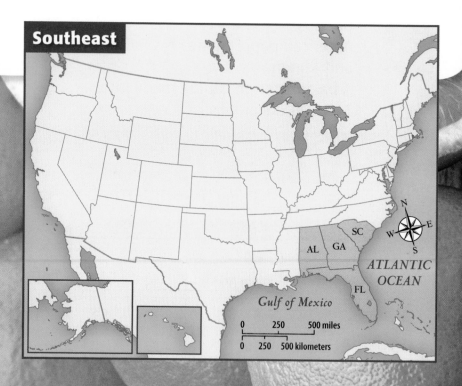

Southeast

SC
AL GA
FL
ATLANTIC OCEAN
Gulf of Mexico

0   250   500 miles
0   250   500 kilometers

Did You Know?

The color of an orange is no indication of its quality because oranges are usually dyed to improve their appearance. Brown spots on the skin indicate a good quality orange.

Florida is the leading agricultural producer of the Southeast region. Florida grows about 4.7 billion pounds of vegetables each year. Look at the chart to see some of this state's top crops.

Florida's Top Crops (2005)		
Crop	Acres Harvested	Yield
Grains and Legumes		
Corn for grain	28,000	94 bushels
Wheat	8,000	45 bushels
Soybeans	8,000	32 bushels
Vegetables		
Potatoes	29,000	13.7 tons
Fruit*		
Citrus	748,600	256.6 million boxes
Other		
Sugarcane	401,000	32.8 tons
Peanuts	152,000	1.4 tons
Cotton	85,000	728 pounds

* The 2005 statistics for oranges are lower than in years before. The hurricanes lowered production of citrus fruit in Florida that year.

The Delta Region

There are 4 big crops grown in the Delta region. Farmers earn the most money from soybeans and cotton. Rice and sugarcane are also important crops for these states.

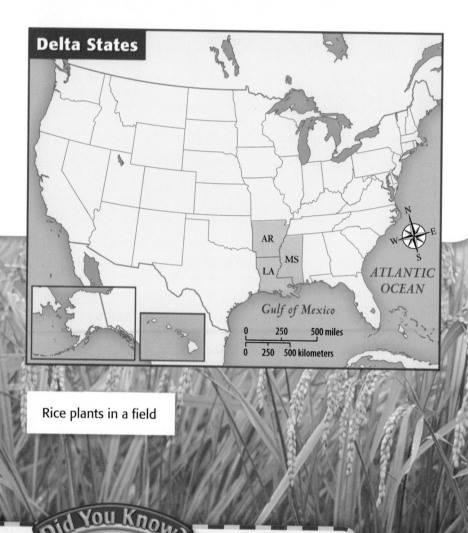

Delta States

AR

MS

LA

ATLANTIC OCEAN

Gulf of Mexico

0   250   500 miles
0   250   500 kilometers

Rice plants in a field

Did You Know?

The lowest part of the sugarcane plant is the sweetest. However, all of the cane can be eaten.

Arkansas is the leading agricultural producer of the Delta region. It ranks 11th in the country for agricultural products. Look at the chart to see some of this state's top crops.

Arkansas' Top Crops (2005)		
Crop	Acres Harvested	Yield
Grains and Legumes		
Soybeans	3,000,000	34 bushels
Rice	1,635,000	332.5 tons
Wheat	160,000	52 bushels
Other		
Cotton	1,040,000	1,011 pounds

Sugarcane field

The Northern Plains Region

The growing season is short in the Northern Plains. The climate and soil are good for growing wheat. This region produces about $\frac{3}{5}$ of winter and spring wheat in the United States.

Corn and sorghum (SOHR guhm) are other grains grown in this region. Sorghum is a tall grass used to feed animals.

Nebraska is the leading agricultural producer of the Northern Plains region. Almost 93% of the land in Nebraska is used for farming. Look at the chart to see some of this state's top crops.

Nebraska's Top Crops (2005)		
Crop	**Acres Harvested**	**Yield**
Grains and Legumes		
Corn	8,250,000	154 bushels
Soybeans	4,660,000	50 bushels
Wheat	1,760,000	39 bushels
Sorghum	250,000	87 bushels
Pinto beans	85,000	1.19 tons
Vegetables		
Potatoes	19,400	21.25 tons

A soybean field in Nebraska.

The Southern Plains Region

There are only two states in the Southern Plains region. A lot of farming takes place in these states. Texas is a leading grower of cotton. Many other crops, such as citrus fruit, nuts, and grains, are also grown in Texas. Oklahoma grows wheat and peanuts.

Southern Plains

OK

TX

ATLANTIC OCEAN

Gulf of Mexico

| 0 | 250 | 500 miles |

| 0 | 250 | 500 kilometers |

Texas is the leading agricultural producer of the Southern Plains region. The average farm in Texas is 575 acres. There are 226,000 farms in Texas. Look at the chart to see some of this state's top crops.

Texas' Top Crops (2005)		
Crop	Acres Harvested	Yield
Grains and Legumes		
Cotton	5,500,000	716 pounds
Hay	5,050,000	1.8 tons
Vegetables		
Potatoes	17,800	17.1 tons
Other		
Peanuts	260,000	1.8 tons
Sunflower seeds	140,000	1,403 pounds

1 ton = 2,000 pounds

The Mountain Region

The state with the most farms per **acre** in this region is Arizona. The land in this region is dry. The states depend on **irrigation** for growing crops. Common products grown include hay, wheat, and some vegetables.

Mountain

MT

ID

WY

NV

UT

CO

AZ

NM

N
W E
S

ATLANTIC
OCEAN

Gulf of Mexico

0   250   500 miles

0   250   500 kilometers

Did You Know?

Sunflowers are native to North America. Early Native Americans used sunflowers for food. They used the oil from sunflowers for their hair.

Colorado is the leading agricultural producer of the Mountain region. It ranks 16th in the entire country. Look at the chart to see some of this state's top crops.

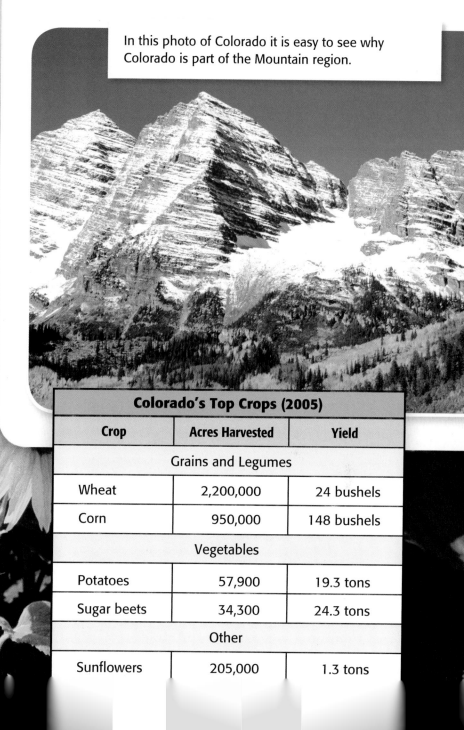

In this photo of Colorado it is easy to see why Colorado is part of the Mountain region.

Colorado's Top Crops (2005)

Crop	Acres Harvested	Yield
Grains and Legumes		
Wheat	2,200,000	24 bushels
Corn	950,000	148 bushels
Vegetables		
Potatoes	57,900	19.3 tons
Sugar beets	34,300	24.3 tons
Other		
Sunflowers	205,000	1.3 tons

The Pacific Region

California leads the nation in growing grapes, lettuce, tomatoes, and strawberries.

More apples are grown in Washington than any other state. The only place where coffee grows in the United States is Hawaii. Hawaii grows the most pineapples and sugarcane.

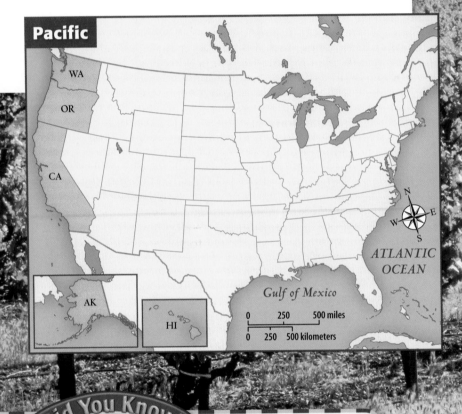

Pacific

WA
OR
CA
AK
HI

ATLANTIC OCEAN

Gulf of Mexico

0   250   500 miles
0   250   500 kilometers

Did You Know?

More than 3,000 California schools have student gardens. Research shows that people who grow their own vegetables tend to like them better.

California grows more food than any other state in the Pacific region or the entire country. Look at the chart to see some of California's important crops.

Some states produce more crops than other states. The United States has many different grains, fruits, and vegetables. What crops grow where you live?

California's Top Crops (2005)		
Crop	Acres Harvested	Yield
Grains and Legumes		
Hay	1,550,000	5.8 tons
Rice	526,000	369.0 tons
Vegetables		
Lettuce	2,230,000	48.0 tons
Sugar beets	44,100	38.7 tons
Fruit		
Grapes	800,000	8.7 tons
Melons	79,900	47.8 tons
Tomatoes	40,000	14.0 tons
Other		
Cotton	657,000	1,191 pounds

Glossary

acre
> A unit of area, equal to 43,560 square feet, used to measure land. *(page 20)*

agriculture
> The business of growing crops and raising animals. *(page 5)*

citrus
> A group of trees grown for their juicy fruit, such as the orange, lemon, and grapefruit. *(page 12)*

irrigation
> The use of ditches or pipes to bring water to fields. *(page 20)*

legumes
> Plants that grow their seeds and fruits in pods, like beans. *(page 6)*